3 MINUTE MATHS

FRACTIONS

KIERAN BALL

Copyright © 2018 Kieran Ball

All rights reserved.

Hello and welcome to 3 Minute Maths – Fractions.

This set of lessons will have you mastering fractions in no time. But don't think that you have to work through this whole book in one go. This book is perfect for the student who only has a few minutes to spare each day. I've set out each lesson so that they can be completed in short bursts, whenever you find yourself with a moment of freedom.

Before you start this book, let me share with you an extremely handy learning tip that helped me to progress quickly and effectively in learning anything. When most people start studying for anything, they tend to spend the first day or two studying for hours and hours and get through loads of work, however, very quickly this begins to dwindle. You might feel like spending hours studying maths, which is great, but you want that feeling of motivation to continue. However, your motivation won't continue if you actually spend hours studying. Limit your study time to chunks of just three minutes. No more!

If you limit your studying to just three-minute chunks, there are three things that will happen...

1. You'll maintain your enthusiasm

If you want to learn anything, you have to maintain enthusiasm or else you won't continue. If you limit your study time to just three minutes, you'll keep maths fresh and exciting and you'll be eager to learn. If you spend hours studying, very quickly you'll get bored with maths and it'll turn into a chore.

2. You'll study more consistently

It's much better to study for just three minutes once a day than to study for three hours once a week. A spare three minutes is relatively easy to find even in the most hectic of schedules. If you make sure you complete at least one three-minute study session every day, it'll quickly become a habit that you'll do without thinking. It's much easier to fit in a daily three-minute habit than a weekly one-hour habit. By doing this, you'll become a much more consistent learner, and consistency is the key to success.

3. You'll remember things better

This is my favourite reason as to why you should limit your study to just three-minute chunks. If you study something for just three minutes every day, you'll trick your brain into memorising the information more quickly than if it were to see the information for hours each day. It'll think, "I see this information every day, so it must be important, but I don't see it for very long, so I'd better hold onto it and make it into a memory fast!" You'll be amazed at how much more easily things tend to stay in your brain if you limit yourself to just three minutes a day.

So, three minutes is the key! Do a three-minute study session and then give yourself at least half an hour before you do another three minutes. You should aim to do at least one three-minute chunk every day, any more than that is a bonus, but one is fine. Just read through this book and complete the exercises, that's all. You'll be amazed at how much you learn.

CONTENTS

LESSON 1: FINDING A FRACTION OF A WHOLE NUMBER

If you were asked to find something like $\frac{4}{9}$ of 18, there is a really simple way to do it. All you have to remember is **"divide by the bottom then times by the top"**.

Say it out loud to yourself fifteen times. Go on!

"Divide by the bottom, times by the top"
"Divide by the bottom, times by the top"
"Divide by the bottom, times by the top"
"Divide by the bottom, times by the top"
"Divide by the bottom, times by the top"
"Divide by the bottom, times by the top"
"Divide by the bottom, times by the top"
"Divide by the bottom, times by the top"
"Divide by the bottom, times by the top"
"Divide by the bottom, times by the top"
"Divide by the bottom, times by the top"
"Divide by the bottom, times by the top"
"Divide by the bottom, times by the top"
"Divide by the bottom, times by the top"
"Divide by the bottom, times by the top"

Now, you're probably asking yourself, "But what does that mean?" Well, hold your horses and I shall tell you.

$$\frac{4}{9}\text{of } 18$$

What you have to do is 18 divided by 9 (the bottom) and then times by 4 (the top)

So, 18 ÷ 9 = 2 and then 2 × 4 = 8

So, $\frac{4}{9}$ of 18 is 8

Easy peasy, lemon squeezy

Let's try another one together. How could we find $\frac{7}{8}$ of 24?

24 ÷ 8 = 3 and then 3 × 7 = 21

So, $\frac{7}{8}$ of 24 is 21

Let's have a quick practice then:

Remember, though, you don't have to do all the questions in one go. Just do a couple every day, because this will help you to retain the knowledge meaning you'll never have to sit and revise for hundreds of hours before your exams.

1. $\frac{3}{7}$ of 35 =
2. $\frac{6}{11}$ of 77 =
3. $\frac{2}{3}$ of 6 =
4. $\frac{8}{9}$ of 36 =
5. $\frac{2}{9}$ of 18 =
6. $\frac{1}{5}$ of 90 =
7. $\frac{3}{4}$ of 92 =
8. $\frac{5}{6}$ of 48 =
9. $\frac{4}{5}$ of 20 =
10. $\frac{3}{7}$ of 21 =

Answers:

1. 15 *(35 ÷ 7 = 5 and 5 × 3 = 15)*
2. 42 *(77 ÷ 11 = 7 and 7 × 6 = 42)*
3. 4 *(6 ÷ 3 = 2 and 2 × 2 = 4)*
4. 32 *(36 ÷ 9 = 4 and 4 × 8 = 32)*
5. 4 *(18 ÷ 9 = 2 and 2 × 2 = 4)*
6. 18 *(90 ÷ 5 = 18 and 18 × 1 = 18)*
7. 69 *(92 ÷ 4 = 23 and 23 × 3 = 69)*
8. 40 *(48 ÷ 6 = 8 and 8 × 5 = 40)*
9. 16 *(20 ÷ 5 = 4 and 4 × 4 = 16)*
10. 9 *(21 ÷ 7 = 3 and 3 × 3 = 9)*

Did you get all of those right? Well done if you did 😊

ADDING, SUBTRACTING, MULTIPLYING & DIVIDING FRACTIONS

These four things are relatively simple things to do with fractions, but let's start with the easiest one: multiplying fractions.

LESSON 2: MULTIPLYING FRACTIONS

The top of a fraction is called the **numerator** and the bottom of a fraction is called the **denominator**. To multiply two fractions together, all you have to do is times the tops together and then times the bottoms together.

For example, if you had $\frac{2}{3} \times \frac{4}{5}$

You do 2 × 4 on top, and then 3 × 5 on the bottom, and you get $\frac{8}{15}$

So, $\frac{2}{3} \times \frac{4}{5} = \frac{8}{15}$

Let's try another one: $\frac{2}{5} \times \frac{3}{7}$

Well, 2 × 3 = 6 and 5 × 7 = 35

So, $\frac{2}{5} \times \frac{3}{7} = \frac{6}{35}$

Let's practise:

1. $\frac{4}{9} \times \frac{3}{4} =$

2. $\frac{1}{2} \times \frac{8}{9} =$

3. $\frac{6}{7} \times \frac{2}{3} =$

4. $\frac{2}{3} \times \frac{3}{4} =$

5. $\frac{2}{5} \times \frac{3}{5} =$

6. $\frac{4}{11} \times \frac{5}{7} =$

7. $\frac{8}{9} \times \frac{3}{10} =$

8. $\frac{3}{8} \times \frac{1}{2} =$

9. $\frac{4}{7} \times \frac{1}{3} =$

10. $\frac{1}{9} \times \frac{7}{8} =$

Answers:

1. $\frac{12}{36}$
2. $\frac{8}{18}$
3. $\frac{12}{21}$
4. $\frac{6}{12}$
5. $\frac{6}{25}$
6. $\frac{20}{77}$
7. $\frac{24}{90}$
8. $\frac{3}{16}$
9. $\frac{4}{21}$
10. $\frac{7}{72}$

SIMPLIFYING FRACTIONS

Now, there were a couple of answers to those questions that weren't in what we call their "simplest form". For example, if you look back at question 1

$$\frac{4}{9} \times \frac{3}{4} = \frac{12}{36}$$

The answer $\frac{12}{36}$ is perfectly fine, but if you are ever asked to give your answer in its simplest form, then you will have to do an extra thing.

12 and 36 are both even numbers, which means you can halve them. If you take a fraction and you halve the top and the bottom, you're simplifying it. So, $\frac{12}{36}$ becomes $\frac{6}{18}$

$\frac{6}{18}$ is a simpler version of $\frac{12}{36}$ but it's the same fraction.

In fact, because 6 and 18 are both even numbers, we could go ahead and halve them again to get $\frac{3}{9}$

$\frac{3}{9}$ is a simpler version of $\frac{6}{18}$ and $\frac{12}{36}$

What we could also do is simplify $\frac{3}{9}$. Both 3 and 9 are in the three times table, which means we could divide them both by 3:

$3 \div 3 = 1$ and $9 \div 3 = 3$, so we could get $\frac{1}{3}$

$\frac{1}{3}$ is a simplified version of $\frac{3}{9}$, $\frac{6}{18}$ and $\frac{12}{36}$

So, we halved it twice and then divided it by 3. We could have done it in one step, by dividing the 12 and the 36 in $\frac{12}{36}$ by 12.

$12 \div 12 = 1$ and $36 \div 12 = 3$ meaning we would get $\frac{1}{3}$, which is the same as what we got before. So, you can simplify any fraction by seeing what you could divide the top and bottom by. I always just try and halve it first if it's an even number, and then see if there's another number you could divide it by, to make it simpler.

Let's try another two. If you look at questions 2 and 4, we can simplify them.

We had $\frac{1}{2} \times \frac{8}{9} = \frac{8}{18}$

Well, again, both the numerator and the denominator (the top number and the bottom number) are even, meaning we can halve them and get $\frac{4}{9}$.

$\frac{4}{9}$ is a simpler version of $\frac{8}{18}$

There's no other number that we can divide both 4 and 9 by, so $\frac{4}{9}$ is the simplest that we can get.

Then, in question 4, we had the question:

$$\frac{2}{3} \times \frac{3}{4} = \frac{6}{12}$$

Both 6 and 12 are even, so let's halve them: $\frac{3}{6}$

$\frac{3}{6}$ is a simpler version of $\frac{6}{12}$

Both 3 and 6 are in the 3 times table, so we can divide them both by 3 and get $\frac{1}{2}$

So, the simplest version of $\frac{6}{12}$ is $\frac{1}{2}$

Let's try one more

What would be $\frac{3}{4} \times \frac{5}{12}$?

$3 \times 5 = 15$ and $4 \times 12 = 48$ so $\frac{3}{4} \times \frac{5}{12} = \frac{15}{48}$

When you look at $\frac{15}{48}$ straight away, you can see that the top isn't even, so we can't halve it. However, we can do something else to them. 15 and 48 are both in the 3 times tables, so what we can do is divide them both by 3

$\frac{15}{48}$ would become $\frac{5}{16}$, which is the simplest version of that fraction.

"But how did you know that 48 was in the 3 times table?" I hear you scream. Well, let me show you a little trick...

THE 3 TIMES TABLE

If you add all the digits in a number together, you can work out whether or not it's in the 3 times table, and therefore whether or not you can divide it by 3.

For example, 48

4 + 8 = 12
and then do the same thing with 12
1 + 2 = 3
This means that 48 is divisible by 3

So, add all the digits in a number together, and then do the same again with the answer until you get down to a single digit number. If you end up with either a 3, a 6 or a 9, then you can divide the original number by 3. How easy is that?!

Which of these numbers can you divide by 3?

1. 4,298
2. 1,231
3. 9,830,283
4. 23,192
5. 894
6. 340,873,452

Answers

1. No *(4 + 2 + 9 + 8 = 23 and then 2 + 3 = 5)*
2. No *(1 + 2 + 3 + 1 = 7)*
3. Yes *(9 + 8 + 3 + 0 + 2 + 8 + 3 = 33 and then 3 + 3 = 6)*
4. No *(2 + 3 + 1 + 9 + 2 = 17 and then 1 + 7 = 8)*
5. Yes *(8 + 9 + 4 = 21 and then 2 + 1 = 3)*
6. Yes *(3 + 4 + 0 + 8 + 7 + 3 + 4 + 5 + 2 = 36 and then 3 + 6 = 9)*

There are some things to look out for 👀 that will make it easier for you to simplify a fraction:

What to look out for to simplify fractions	
Even number	*You can divide it by 2*
Add the digits together and get 3, 6 or 9	*You can divide it by 3*
End in a 0	*You can divide it by 10*
End in a 0 or a 5	*You can divide it by 5*
Double digit numbers	*You can divide it by 11*

So, using those little nuggets of wisdom, see if you can simplify the following fractions:

1. $\frac{415}{985}$
2. $\frac{322}{704}$
3. $\frac{99}{120}$
4. $\frac{66}{88}$
5. $\frac{72}{102}$

Answers

1. $\frac{83}{197}$ *(÷5)*
2. $\frac{161}{352}$ *(÷2)*
3. $\frac{33}{40}$ *(÷3)*
4. $\frac{3}{4}$ *(÷11 = 6/8 and then ÷2)*
5. $\frac{12}{17}$ *(÷2 = 36/51 and then ÷3)*

Anyway, let's get back to what we were doing. So, now you know how to multiply fractions; you just multiply the numerators together, and then you multiply the denominators together. But what about if you wanted to multiply a fraction with a normal whole number?

Well, any normal whole number can be turned into a fraction extremely easily. All you have to do is place it over a denominator of 1. For example, 15 as a fraction is $\frac{15}{1}$.

So, if you want to do something like $\frac{4}{5} \times 12$ you start by turning 12 into a fraction:

$$\frac{4}{5} \times \frac{12}{1} = \frac{48}{5}$$

Or what about this one: $8 \times \frac{2}{3}$

Well, you change 8 into a fraction, and you get $\frac{8}{1} \times \frac{2}{3} = \frac{16}{3}$

Have a go at doing these five questions:

1. $4\times\frac{3}{5}=$
2. $\frac{2}{3}\times7=$
3. $8\times\frac{1}{10}=$
4. $\frac{3}{7}\times8=$
5. $4\times\frac{1}{2}=$

Answers

1. $\frac{12}{5}$ $\left(\frac{4}{1}\times\frac{3}{5}\right)$
2. $\frac{14}{3}$ $\left(\frac{2}{3}\times\frac{7}{1}\right)$
3. $\frac{4}{5}$ $\left(\frac{8}{1}\times\frac{1}{10}=\frac{8}{10}\text{ and then divide by }2\right)$
4. $\frac{24}{7}$ $\left(\frac{3}{7}\times\frac{8}{1}\right)$
5. $\frac{2}{1}$ $(\frac{4}{1}\times\frac{1}{2}=\frac{4}{2}\text{ and then divide by }2)$

Another challenge complete ☑

Let's move on to the next easiest challenge: dividing fractions

LESSON 3: DIVIDING FRACTIONS

Dividing fractions is almost as easy as multiplying fractions, but there's one extra little step.

Let's look at $\frac{2}{3} \div \frac{5}{6}$

Now, there are two steps:
STEP 1: Flip the second fraction
STEP 2: Multiply

So, you have to turn the second fraction upside down, so $\frac{5}{6}$ would become $\frac{6}{5}$.
Then, you just multiply:

$$\frac{2}{3} \times \frac{6}{5} = \frac{12}{15}$$

Now, you can also simplify $\frac{12}{15}$, if you want. If you don't want to, look away and ignore the next few lines.

I said don't look!

Have you gone?

Anyway, for those of you who wanted to simplify $\frac{12}{15}$, here it is...

They're both in the 3 times table, so we can divide the top and bottom by 3 and get $\frac{4}{5}$

Phew.

Let's do a couple more practice questions together:

$$\frac{3}{7} \div \frac{1}{2} =$$

Remember, to flip the second fraction over, and then multiply:

$$\frac{3}{7} \times \frac{2}{1} = \frac{6}{7}$$

We can't simplify this fraction; it's as simple as can be.

Let's do one more (hint hint, it's going to lead on to something else ☺)

$$\frac{4}{5} \div \frac{2}{3} =$$

Flip and multiply!

$$\frac{4}{5} \times \frac{3}{2} = \frac{12}{10}$$

Now, now, look what we have here. Whenever you see a fraction where the numerator is bigger than the denominator, it's called an **improper fraction**, because it's just not proper! You could also call it a **top-heavy fraction**.

TURNING IMPROPER FRACTIONS INTO MIXED FRACTIONS

Whenever you get an **improper fraction**, you can do something marvellous with it: make it into a **mixed fraction**!

So, let's start by just simplifying our improper fraction, $\frac{12}{10}$, to make it as small as we can get it so it's easier to handle. The numerator and the denominator are both even, meaning we can halve them and get $\frac{6}{5}$

$\frac{6}{5}$ is still an improper fraction, it's just a simpler one. Here's how to turn an improper fraction into a mixed fraction:

STEP 1: Work out how many times the denominator can fit into the numerator and write it as a big number to the left
STEP 2: Whatever is left over goes as the new numerator

$$\frac{6}{5}$$

5 can fit into 6 just 1 time, so we write a big 1. Then, there is 1 left over, so the mixed fraction is $1\frac{1}{5}$

Let's do another so you can get the pattern:

$\frac{17}{5}$ is an improper fraction, let's make it into a mixed fraction:

How many times does 5 fit into 17? 3 times with 2 left over.

So, we write a big 3 and then make 2 the new numerator: $3\frac{2}{5}$

Let's do two more:

$$\frac{29}{6}$$

How many times does 6 fit into 29? 4 times with 5 left over.

So, we write a big 4 and then make 5 the new numerator: $4\frac{5}{6}$

What about this one? $\frac{56}{32}$

This one can be simplified first to make it easier. Since both numbers are even, we can halve them: $\frac{28}{16}$

We could halve them again because 28 and 16 are both even too: $\frac{14}{8}$

And we could halve them a third time: $\frac{7}{4}$

Now, turn the improper fraction $\frac{7}{4}$ into a mixed fraction.

How many times does 4 fit into 7? Just 1 time with 3 left over

So, we write a big 1 and then the new numerator is 3: $1\frac{3}{4}$

Now, for this one, I said we should simplify the fraction first, but what if you don't want to? Well, good news because you don't have to because you can just simplify at the end instead.

Let's do the same question again, but let's not simplify it first:

$$\frac{56}{32}$$

How many times does 32 fit into 56? Just 1 time with 24 left over.
So, we can write a big 1 and then the new numerator is 24: $1\frac{24}{32}$

We can now just simplify the fraction part. Both 24 and 32 are even, so we can halve them both: $\frac{12}{16}$

We can halve them again because 12 and 16 are both even: $\frac{6}{8}$

We can even halve them a third time: $\frac{3}{4}$

If we put the big 1 back, we get $1\frac{3}{4}$, which is exactly the same as we got by simplifying the fraction at the start. So, you can either simplify the fraction at the beginning or simplify it at the end, it doesn't matter.

There's one more type of improper fraction that you might come across. Something like this: $\frac{20}{5}$

So, how many times does 5 fit into 20? 4 with none left over. This means the answer is just 4

If we had simplified this first, you could divide the top and bottom by 5 to get $\frac{4}{1}$. Anytime you have a fraction with 1 as the denominator, you can just ignore the denominator, so the answer is just 4.

Have a go at turning these improper fractions into mixed fractions (you might want to simplify them a little first, but remember that you don't have to because you can also simplify them at the end instead):

1. $\frac{19}{3}$
2. $\frac{45}{7}$
3. $\frac{16}{5}$
4. $\frac{23}{12}$
5. $\frac{27}{12}$
6. $\frac{108}{56}$
7. $\frac{240}{32}$
8. $\frac{500}{200}$
9. $\frac{99}{33}$
10. $\frac{3}{2}$

Answers

1. $6\frac{1}{3}$ (How many times does 3 fit into 19? 6 with 1 left over)

2. $6\frac{3}{7}$ (How many times does 7 fit into 45? 6 with 3 left over)

3. $3\frac{1}{5}$ (How many times does 5 fit into 16? 3 with 1 left over)

4. $1\frac{11}{12}$ (How many times does 12 fit into 23? 1 with 11 left over)

5. $2\frac{1}{4}$ (If you simplify first, you can ÷3 and get $\frac{9}{4}$. Then how many times does 4 fit into 9? 2 with 1 left over.
If you don't simplify first you say how many times does 12 fit into 27? 2 with 3 left over getting $2\frac{3}{12}$. Then you can simplify the $\frac{3}{12}$ by dividing both by 3, getting $\frac{1}{4}$)

6. $1\frac{13}{14}$ (If you simplify first, you can halve it to get $\frac{54}{28}$ and then halve it again to get $\frac{27}{14}$. How many times does 14 fit into 27? 1 with 13 left over.
If you don't simplify first, you say how many times does 56 fit into 108? 1 with 52 left over giving $1\frac{52}{56}$. You can then simplify the $\frac{52}{56}$ by ÷2 to get $\frac{26}{28}$ and then ÷2 again to get $\frac{13}{14}$)

7. $7\frac{1}{2}$ (If you simplify first, you can halve them to get $\frac{120}{16}$ and then halve again to get $\frac{60}{8}$ and then halve again to get $\frac{30}{4}$ and then halve again to get $\frac{15}{2}$. How many times does 2 fit into 15? 7 with 1 left over.
If you didn't want to simplify first, you could ask how many times 32 fits into 240. 7 with 16 left over giving you $7\frac{16}{32}$. You can then simplify $\frac{16}{32}$ by halving to get $\frac{8}{16}$, halving again to get $\frac{4}{8}$ and then halving two more times to get $\frac{2}{4}$ and then $\frac{1}{2}$)

8. $2\frac{1}{2}$ (If you simplify them first, you could divide by 100 and get $\frac{5}{2}$. How many times does 2 fit into 5? 2 with 1 left over.
If you didn't simplify first, you say how many times does 200 fit into 500? 2 with 100 left over, giving you $2\frac{100}{200}$. You can then simplify $\frac{100}{200}$ by dividing the top and the bottom by 100 to get $\frac{1}{2}$)

9. 3 (If you simplify first, you can divide by 11 and get $\frac{9}{3}$ and then divide by 3 to get $\frac{3}{1}$.
Whenever you have a fraction over 1, you can just ignore the denominator, so the answer is 3.
If you didn't simplify first, you could ask, how many times does 33 fit into 99? 3 times exactly)

10. $1\frac{1}{2}$ (How many times does 2 fit into 3? 1 with 1 left over)

So, that's improper fractions being turned into mixed fractions, but you could also do it the other way around and turn a mixed fraction into an improper fraction. Read on and find out how...

TURNING MIXED FRACTIONS INTO IMPROPER FRACTIONS

Let's say you start out with a mixed fraction like $2\frac{5}{7}$
If you want to turn this into an improper fraction, all you have to do is do 2 × 7 to get 14, and then add it onto the numerator. 5 + 14 = 19, giving you $\frac{19}{7}$

Let's try another. Turn $4\frac{8}{11}$ into an improper fraction.
4 × 11 = 44 and then 44 + 8 = 52. So, you get $\frac{52}{11}$

If it's possible, you could simplify the fraction bit first. For example: $3\frac{12}{15}$
You can simplify $\frac{12}{15}$ by dividing the top and bottom by 3, giving you $\frac{4}{5}$
Then you can turn $3\frac{1}{5}$ into an improper fraction.
3 × 5 = 15 and then 15 + 4 = 19, giving you $\frac{19}{5}$

If you didn't simplify it first, you could still do it, but you would have a slightly trickier multiplication.

$$3\frac{12}{15}$$

3 × 15 = 45 and then 45 + 12 = 57 giving you $\frac{57}{15}$

Both the numerator and denominator are divisible by 3, so you would get $\frac{19}{5}$

Here are some practice questions for you to practise with. Turn these mixed fractions into improper fractions:

1. $5\frac{2}{3}$
2. $6\frac{7}{10}$
3. $20\frac{3}{8}$
4. $4\frac{1}{4}$
5. $5\frac{1}{2}$

6. $9\frac{1}{10}$

7. $3\frac{1}{12}$

8. $4\frac{6}{7}$

9. $10\frac{11}{22}$

10. $10\frac{4}{6}$

Answers:

1. $\frac{17}{3}$ (5 × 3 = 15 and then 15 + 2 =17)

2. $\frac{67}{10}$ (6 × 10 = 60 and then 60 + 7 = 67)

3. $\frac{163}{8}$ (20 × 8 = 160 and then 160 + 3 = 163)

4. $\frac{17}{4}$ (4 × 4 = 16 and then 16 + 1 = 17)

5. $\frac{11}{2}$ (5 × 2 = 10 and then 10 + 1 = 11)

6. $\frac{91}{10}$ (9 × 10 = 90 and then 90 + 1 = 91)

7. $\frac{37}{12}$ (3 × 12 = 36 and then 36 + 1 = 37)

8. $\frac{34}{7}$ (4 × 7 = 28 and then 28 + 6 = 34)

9. $\frac{21}{2}$ (If you simplify the fraction at the start by dividing it by 11, you get $\frac{1}{2}$ and then 10 × 2 = 20 and 20 + 1 = 21.
If you didn't simplify it first, you would have done 10 × 22 = 220 and then 220 + 11 = 231. You would then have $\frac{231}{22}$, which you could ÷11 to get $\frac{21}{2}$)

10. $\frac{32}{3}$ (If you simplified the fraction first, you could have halved the numerator and the denominator, getting $\frac{2}{3}$. Then 10 × 3 = 30 and 30 + 2 = 32.
Or, by not simplifying first, you would have done 10 × 6 = 60 and then 60 + 4 = 64, giving you $\frac{64}{6}$. You could halve this and get $\frac{32}{3}$)

Goodness me, we've done a lot about improper fractions and mixed fractions, now let's get back to dividing fractions.

DIVIDING FRACTIONS continued...

Remember that when you divide fractions, you have to start by flipping over the second fraction, and then you multiply. So, have a go at these ten questions at your leisure:

1. $\frac{4}{5} \div \frac{8}{9}$
2. $\frac{3}{7} \div \frac{9}{11}$
3. $\frac{1}{2} \div \frac{5}{9}$
4. $\frac{6}{13} \div \frac{2}{3}$
5. $\frac{8}{9} \div \frac{1}{10}$
6. $\frac{6}{11} \div \frac{1}{3}$
7. $\frac{1}{3} \div \frac{1}{4}$
8. $\frac{1}{4} \div \frac{1}{2}$
9. $\frac{1}{2} \div \frac{1}{4}$
10. $\frac{1}{8} \div \frac{2}{7}$

Answers

1. $\frac{9}{10}$ ($\frac{4}{5} \times \frac{9}{8} = \frac{36}{40}$ halve it to get $\frac{18}{20}$ and halve again to get $\frac{9}{10}$)
2. $\frac{11}{21}$ ($\frac{3}{7} \times \frac{11}{9} = \frac{33}{63}$ and then ÷3 to simplify into $\frac{11}{21}$)
3. $\frac{9}{10}$ ($\frac{1}{2} \times \frac{9}{5} = \frac{9}{10}$)
4. $\frac{9}{13}$ ($\frac{6}{13} \times \frac{3}{2} = \frac{18}{26}$ and then halve it to get $\frac{9}{13}$)

5. $8\frac{8}{9}$ ($\frac{8}{9} \times \frac{10}{1} = \frac{80}{9}$, which you could make into a mixed fraction. How many times does 9 go into 80? 8 times with 8 left over)

6. $1\frac{7}{11}$ ($\frac{6}{11} \times \frac{3}{1} = \frac{18}{11}$ and you can mix it up fraction style by saying, how many times does 11 go into 18? 1 time with 7 left over)

7. $1\frac{1}{3}$ ($\frac{1}{3} \times \frac{4}{1} = \frac{4}{3}$ and then make it mixed by asking how many times 3 goes into 4. 1 time with 1 left over)

8. $\frac{1}{2}$ ($\frac{1}{4} \times \frac{2}{1} = \frac{2}{4}$. If you halve the numerator and the denominator, you get $\frac{1}{2}$)

9. 2 ($\frac{1}{2} \times \frac{4}{1} = \frac{4}{2}$. How many times does 2 go into 4? 2 times exactly)

10. $\frac{7}{16}$ ($\frac{1}{8} \times \frac{7}{2} = \frac{7}{16}$)

But wait! Wait! WAAAIIIIIITTTT!!!

What about if we just want to do a normal whole number divided by a fraction? Or a fraction divided by a normal whole number? Well, it's easy. Let's see...

Remember how I said that any whole number can be made into a fraction simply by placing it over a denominator of 1? Well, good, because we'll need that factoid now.
If you had to work out $\frac{3}{4} \div 4$, then what you should start by doing is turning 4 into $\frac{4}{1}$

So, then you would have $\frac{3}{4} \div \frac{4}{1}$ and you can do it normally by flipping the second fraction over and multiplying

$$\frac{3}{4} \times \frac{1}{4} = \frac{3}{16}$$

Similarly, if you want to take a whole number and divide it by a fraction, you can just turn the whole number into a fraction and do as normal.
For example: $7 \div \frac{2}{3}$, start by turning the 7 into $\frac{7}{1}$

$$\frac{7}{1} \div \frac{2}{3} =$$

$$\frac{7}{1} \times \frac{3}{2} = \frac{21}{2}$$

As a mixed fraction, that would be $10\frac{1}{2}$

So, have a go at these five questions:

1. $5 \div \frac{4}{7}$
2. $\frac{8}{9} \div 4$
3. $\frac{6}{11} \div 8$
4. $10 \div \frac{2}{5}$
5. $\frac{2}{3} \div 2$

Answers

1. $8\frac{3}{4}$ ($\frac{5}{1} \times \frac{7}{4} = \frac{35}{4}$, How many times does 4 go into 35? 8 with 3 left over)
2. $\frac{2}{9}$ ($\frac{8}{9} \times \frac{1}{4} = \frac{8}{36}$ and we can divide the top and bottom by 4 to get $\frac{2}{9}$)
3. $\frac{3}{44}$ ($\frac{6}{11} \times \frac{1}{8} = \frac{6}{88}$ and then divide by 2)
4. 25 ($\frac{10}{1} \times \frac{5}{2} = \frac{50}{2}$. How many times does 2 go into 50? 25 times exactly)
5. $\frac{1}{3}$ ($\frac{2}{3} \times \frac{1}{2} = \frac{2}{6}$ and then halve the numerator and the denominator)

So, now we know that to multiply fractions, you just times the numerators and then times the denominators. We also know that to divide fractions, you flip the second fraction upside down, and then multiply.

Now, let's move on to adding fractions.

LESSON 4: ADDING FRACTIONS

Adding two fractions together is slightly trickier than multiplying and dividing was, but it's not too bad. All you have to remember is that the denominators have to be the same.

The reason the denominators have to be the same is so that they are both the same sort of fraction. It's the same as if you were to have 20 euros in one hand and 30 pounds in the other. You couldn't just add the 20 and 30 together and say you have 50 euros or 50 pounds, because they're not the same currency. However, if you went and exchanged your euros for pounds, then you could add them altogether.

So, let's have a look at how to make sure the denominators are the same:

$$\frac{2}{3} + \frac{5}{6} =$$

If you look at these two fractions, the denominators are different, but you can make them the same. If you times the top and bottom of $\frac{2}{3}$ by 2, you'll get $\frac{4}{6}$. Now you have two fractions with the same denominators:

$$\frac{4}{6} + \frac{5}{6} = \frac{9}{6}$$

Once you have the same denominators, you just add the numerators together.

Let's try another one:

$$\frac{3}{8} + \frac{1}{4} =$$

Well, if you take the $\frac{1}{4}$ and multiply the numerator and the denominator by 2, you get $\frac{2}{8}$, which has the same denominator as the other fraction. Put it back into the equation and add the numerators.

$$\frac{3}{8} + \frac{2}{8} = \frac{5}{8}$$

What about this one?

$$\frac{1}{5} + \frac{4}{15} =$$

Well, one denominator is 5 and the other is 15. 5 can be turned into 15 by multiplying by 3. So, if you multiply the top and bottom of $\frac{1}{5}$ by 3, you'll get $\frac{3}{15}$. And now you can add them together:

$$\frac{3}{15} + \frac{4}{15} = \frac{7}{15}$$

So far, we've just had to change one denominator so it's the same as the other. But what about if you can't easily make one denominator into the other?

$$\frac{3}{4} + \frac{2}{3} =$$

The denominators are 4 and 3, and you can't make 3 into 4 by multiplying it by anything. So, instead, we're going to have to change both denominators into a completely new "common denominator" (this means the denominators are the same). When we have to do this, a very easy way to do it is to multiply the first fraction by the denominator of the second fraction, and to multiply the second fraction by the denominator of the first fraction.

$$\frac{3}{4} + \frac{2}{3} =$$

If you multiply the top and bottom of $\frac{3}{4}$ by 3, you get $\frac{9}{12}$

If you multiply the top and bottom of $\frac{2}{3}$ by 4, you get $\frac{8}{12}$

Now they both have the same denominators, so we can just add the numerators together:

$$\frac{9}{12} + \frac{8}{12} = \frac{17}{12} \text{ or } 1\frac{5}{12}$$

This is a little bit tricky, so please don't panic if you don't get it straight away. Let's try another one:

$$\frac{4}{5} + \frac{1}{3} =$$

So, multiply the top and bottom of the first fraction by the denominator of the second fraction, and then multiply the top and bottom of the second fraction by the denominator of the first fraction.

If you multiply the top and bottom of $\frac{4}{5}$ by 3, you get $\frac{12}{15}$

If you multiply the top and bottom of $\frac{1}{3}$ by 5, you get $\frac{5}{15}$

Woohoo! They both have the same denominator now, so we can just add the numerators.

$$\frac{12}{15} + \frac{5}{15} = \frac{17}{15} \text{ or } 1\frac{2}{15}$$

We'll work through one more together, and then you can do some practice questions:

$$\frac{3}{10} + \frac{2}{7} =$$

If you multiply the top and bottom of $\frac{3}{10}$ by 7, you get $\frac{21}{70}$

If you multiply the top and bottom of $\frac{2}{7}$ by 10, you get $\frac{20}{70}$

$$\frac{21}{70} + \frac{20}{70} = \frac{41}{70}$$

Have a go at adding these fractions together:

1. $\frac{2}{5} + \frac{1}{6} =$
2. $\frac{3}{4} + \frac{1}{7} =$
3. $\frac{7}{8} + \frac{4}{7} =$
4. $\frac{2}{9} + \frac{2}{3} =$
5. $\frac{3}{7} + \frac{1}{10} =$
6. $\frac{2}{3} + \frac{2}{10} =$
7. $\frac{5}{9} + \frac{2}{7} =$
8. $\frac{1}{8} + \frac{3}{11} =$
9. $\frac{6}{7} + \frac{5}{12} =$
10. $\frac{4}{5} + \frac{1}{3} =$

Answers

1. $\frac{17}{30}$ $(\frac{12}{30}+\frac{5}{30})$
2. $\frac{25}{28}$ $(\frac{21}{28}+\frac{4}{28})$
3. $1\frac{25}{56}$ $(\frac{49}{56}+\frac{32}{56}=\frac{81}{56}$. How many times does 56 go into 81? 1 time with 25 left over)
4. $\frac{8}{9}$ $(\frac{6}{27}+\frac{18}{27}=\frac{24}{27}$, which you can divide by 3)
5. $\frac{37}{70}$ $(\frac{30}{70}+\frac{7}{70})$
6. $\frac{13}{15}$ $(\frac{20}{30}+\frac{6}{30}=\frac{26}{30}$, which you can divide by 2)
7. $\frac{53}{63}$ $(\frac{35}{63}+\frac{18}{63})$
8. $\frac{35}{88}$ $(\frac{11}{88}+\frac{24}{88})$
9. $1\frac{23}{84}$ $(\frac{72}{84}+\frac{35}{84}=\frac{107}{84}$. How many times does 84 go into 107? 1 time with 23 left over)
10. $1\frac{2}{15}$ $(\frac{12}{15}+\frac{5}{15}=\frac{17}{15}$. How many times does 15 go into 17? 1 time with 2 left over)

What about if you want to add fractions to whole numbers? Well, this is quite easy too, and there are two ways you can do it. Firstly, you could just make it into a mixed fraction by shoving the whole number and the fraction next to each other:

$$\frac{2}{3}+5=5\frac{2}{3}$$

$$4+\frac{1}{7}=4\frac{1}{7}$$

Or, you can make the whole number into a fraction by placing it over a denominator of 1.

$\frac{2}{3}+5$ would become $\frac{2}{3}+\frac{5}{1}$ and then you multiply the top and bottom of the second fraction by 3 to get $\frac{15}{3}$:

$$\frac{2}{3}+\frac{15}{3}=\frac{17}{3}$$

Or if we had $4 + \frac{1}{7}$, we can turn it into $\frac{4}{1} + \frac{1}{7}$ and multiply the top and bottom of the first fraction by 7 to get $\frac{28}{7}$

$$\frac{28}{7} + \frac{1}{7} = \frac{29}{7}$$

So, when you're adding fractions together, you need to make sure the denominators are the same. It's the exact same concept when you're subtracting fractions.

LESSON 5: SUBTRACTING FRACTIONS

When you subtract fractions, you have to make sure that the denominators are the same. If you can see an easy way to change one of the fractions so that it has the same denominator as the other fraction, you can just do that. If not, use the same technique as we used when adding the fractions. Multiply the top and bottom of the first fraction by the denominator of the second fraction, and vice versa.

So, let's have a look at an example:

$$\frac{5}{6} - \frac{2}{3} =$$

If you look at these two fractions, the denominators are different, but you can make them the same. If you multiply the top and bottom of $\frac{2}{3}$ by 2, you'll get $\frac{4}{6}$. Now you have two fractions with the same denominators:

$$\frac{5}{6} - \frac{4}{6} = \frac{1}{6}$$

Once you have the same denominators, you just subtract the numerators.

Let's try another one:

$$\frac{3}{8} - \frac{1}{4} =$$

Well, if you take the $\frac{1}{4}$ and multiply the numerator and the denominator by 2, you get $\frac{2}{8}$, which has the same denominator as the other fraction. Put it back into the equation and subtract the numerators.

$$\frac{3}{8} - \frac{2}{8} = \frac{1}{8}$$

What about this one?

$$\frac{4}{15} - \frac{1}{5} =$$

Well, one denominator is 5 and the other is 15. 5 can be turned into 15 by multiplying by 3. So, if you multiply the top and bottom of $\frac{1}{5}$ by 3, you'll get $\frac{3}{15}$. And now you can subtract them:

$$\frac{4}{15} - \frac{3}{15} = \frac{1}{15}$$

What about this one? It's not so easy to change just one denominator to make it look like the other this time.

$$\frac{3}{4} - \frac{2}{3} =$$

If you multiply the top and bottom of $\frac{3}{4}$ by 3, you get $\frac{9}{12}$

If you multiply the top and bottom of $\frac{2}{3}$ by 4, you get $\frac{8}{12}$

Now they both have the same denominators, so we can just subtract the numerators:

$$\frac{9}{12} - \frac{8}{12} = \frac{1}{12}$$

Let's try another one:

$$\frac{4}{5} - \frac{1}{3} =$$

So, multiply the top and bottom of the first fraction by the denominator of the second fraction, and then multiply the top and bottom of the second fraction by the denominator of the first fraction.

If you multiply the top and bottom of $\frac{4}{5}$ by 3, you get $\frac{12}{15}$

If you multiply the top and bottom of $\frac{1}{3}$ by 5, you get $\frac{5}{15}$

$$\frac{12}{15} - \frac{5}{15} = \frac{7}{15}$$

We'll work through one more together, and then you can do some practice questions:

$$\frac{3}{10} - \frac{2}{7} =$$

If you multiply the top and bottom of $\frac{3}{10}$ by 7, you get $\frac{21}{70}$

If you multiply the top and bottom of $\frac{2}{7}$ by 10, you get $\frac{20}{70}$

$$\frac{21}{70} - \frac{20}{70} = \frac{1}{70}$$

Have a go at subtracting these fractions:

1. $\frac{2}{5} - \frac{1}{6} =$
2. $\frac{3}{4} - \frac{1}{7} =$
3. $\frac{7}{8} - \frac{4}{7} =$
4. $\frac{2}{3} - \frac{2}{9} =$
5. $\frac{3}{7} - \frac{1}{10} =$
6. $\frac{2}{3} - \frac{2}{10} =$
7. $\frac{5}{9} - \frac{2}{7} =$
8. $\frac{3}{11} - \frac{1}{8} =$
9. $\frac{6}{7} - \frac{5}{12} =$
10. $\frac{4}{5} - \frac{1}{3} =$

Answers

1. $\frac{7}{30}$ $(\frac{12}{30} - \frac{5}{30})$
2. $\frac{17}{28}$ $(\frac{21}{28} - \frac{4}{28})$
3. $\frac{17}{56}$ $(\frac{49}{56} - \frac{32}{56})$

4. $\frac{4}{9}$ $(\frac{18}{27} - \frac{6}{27} = \frac{12}{27}$ which you can divide by 3)

5. $\frac{23}{70}$ $(\frac{30}{70} - \frac{7}{70})$

6. $\frac{7}{15}$ $(\frac{20}{30} - \frac{6}{30} = \frac{14}{30}$ which you can divide by 2)

7. $\frac{17}{63}$ $(\frac{35}{63} - \frac{18}{63})$

8. $\frac{13}{88}$ $(\frac{24}{88} - \frac{11}{88})$

9. $\frac{37}{84}$ $(\frac{72}{84} - \frac{35}{84})$

10. $\frac{7}{15}$ $(\frac{12}{15} - \frac{5}{15})$

What about if you want to subtract a fraction from a whole number? Well, just like before, all you have to do is turn the whole number into a fraction by putting over a denominator of 1.

$$5 - \frac{2}{3}$$

Write it as: $\frac{5}{1} - \frac{2}{3}$ and then multiply the top and bottom of the first fraction by 3.

$\frac{15}{3} - \frac{2}{3} = \frac{13}{3}$. How many times does 3 go into 13? 4 times with one left over:

$$5 - \frac{2}{3} = 4\frac{1}{3}$$

Or how about $7 - \frac{4}{5}$

Well, rewrite it as this: $\frac{7}{1} - \frac{4}{5}$ and then multiply the numerator and the denominator of the first fraction by 5:

$\frac{35}{5} - \frac{4}{5} = \frac{31}{5}$. How many times does 5 go into 31? 6 with 1 left over:

$$7 - \frac{4}{5} = 6\frac{1}{5}$$

Now, let me show you a secret way of doing this. You might have already figured it out, but if you haven't, put your listening ears on...

WHOLE NUMBER SUBTRACT A FRACTION

A fraction is a number less than one, so when you subtract them from a whole number you're not even taking away 1.

Look at this: $8 - \frac{1}{4}$

Well, think about, what would you have left if had 8 whole pizzas and you took a quarter of one of them away? You would have 7 whole pizzas and $\frac{3}{4}$ of another one left.

So, $8 - \frac{1}{4} = 7\frac{3}{4}$

All you have to do to subtract a fraction from a whole number, is take away one, and then add on the rest of the fraction. You just need to work out what fraction is left over, which isn't too tricky to do.

If you had a pizza and took away the following fractions, what would you have left:

1. 1 pizza take away $\frac{2}{3}$
2. 1 pizza take away $\frac{5}{7}$
3. 1 pizza take away $\frac{1}{10}$
4. 1 pizza take away $\frac{5}{12}$
5. 1 pizza take away $\frac{7}{8}$

Here's a tip, work out what you have to add to the numerator to make the denominator.

Answers:

1. $\frac{1}{3}$
2. $\frac{2}{7}$
3. $\frac{9}{10}$

4. $\frac{7}{12}$

5. $\frac{1}{8}$

So, if instead, you had more than one pizza, all you have to do is subtract one, and then add on a fraction.

Try these:

1. 5 pizzas subtract $\frac{3}{5}$
2. 8 pizzas subtract $\frac{1}{9}$
3. 2 pizzas subtract $\frac{7}{12}$
4. $4 - \frac{5}{6} =$ (this is the same as 4 pizzas subtract $\frac{5}{6}$)
5. $8 - \frac{1}{11} =$
6. $12 - \frac{4}{9} =$
7. $20 - \frac{1}{2} =$
8. $19 - \frac{1}{3} =$
9. $4 - \frac{7}{15} =$
10. $3 - \frac{6}{7} =$

Answers:

1. 4 whole pizzas and $\frac{2}{5}$
2. 7 whole pizzas and $\frac{8}{9}$
3. 1 whole pizza and $\frac{5}{12}$
4. $3\frac{1}{6}$

5. $7\frac{10}{11}$

6. $11\frac{5}{9}$

7. $19\frac{1}{2}$

8. $18\frac{2}{3}$

9. $3\frac{8}{15}$

10. $2\frac{1}{7}$

Now, this is just an extra way to work out how to subtract fractions from whole numbers, but if you prefer the other way or you don't quite get this way yet, that's fine; both ways work 😊

Anyway, that's the end.

RECAP

Let's just recap fast:

FRACTIONS OF WHOLE NUMBERS

Divide by the bottom, times by the top

MULTIPLYING FRACTIONS

Multiply the numerators together, and multiply the denominators together

DIVIDING FRACTIONS

Flip the second fraction upside down and then multiply

ADDING FRACTIONS

Make sure the denominators are the same and then add the numerators

SUBTRACTING FRACTIONS

Make sure the denominators are the same and then subtract the numerators

MAKING WHOLE NUMBERS INTO FRACTIONS

Place the number over a denominator of 1

Let's take two fractions: $\frac{2}{3}$ and $\frac{1}{2}$ and we'll multiply them, divide them, add them and subtract them.

MULTIPLYING

$$\frac{2}{3} \times \frac{1}{2} = \frac{2}{6} \rightarrow \frac{1}{3}$$

DIVIDING

$$\frac{2}{3} \div \frac{1}{2} \rightarrow \frac{2}{3} \times \frac{2}{1} = \frac{4}{3} \rightarrow 1\frac{1}{3}$$

ADDING

$$\frac{2}{3} + \frac{1}{2} \rightarrow \frac{4}{6} + \frac{3}{6} = \frac{7}{6} \rightarrow 1\frac{1}{6}$$

SUBTRACTING

$$\frac{2}{3} - \frac{1}{2} \rightarrow \frac{4}{6} - \frac{3}{6} = \frac{1}{6}$$

PRACTICE QUESTIONS

Here's a mixture of questions for you to have a go at. Don't do them all in one go; just do two or three now and again to keep on top of it. That way you won't forget what you've learnt, and you won't have to do lots of questions in one go:

1. $\frac{4}{5}+\frac{5}{8}$
2. $\frac{7}{9}+\frac{7}{13}$
3. $\frac{9}{20}-\frac{4}{18}$
4. $\frac{1}{2}\times 7$
5. $\frac{2}{5}\div\frac{3}{9}$
6. $\frac{1}{12}+\frac{4}{9}$
7. $\frac{3}{10}-\frac{1}{8}$
8. $\frac{5}{7}+\frac{1}{3}$
9. $\frac{2}{3} of\ 33$
10. $9-\frac{3}{10}$
11. $\frac{5}{7}+3$
12. $\frac{4}{5}-\frac{1}{8}$
13. $\frac{7}{19}-\frac{2}{13}$
14. $\frac{9}{20}+\frac{4}{18}$
15. $\frac{1}{2}\div 7$
16. $\frac{2}{5}\times\frac{3}{9}$
17. $\frac{1}{12}-\frac{1}{15}$
18. $\frac{3}{10}+\frac{1}{8}$
19. $\frac{5}{7}-\frac{1}{3}$
20. $\frac{2}{5} of\ 40$
21. $4+\frac{1}{3}$
22. $\frac{2}{5} of\ 15$
23. $5\times\frac{2}{7}$
24. $\frac{4}{9}\times\frac{3}{4}$
25. $\frac{1}{2}\div\frac{3}{5}$
26. $\frac{7}{9}+\frac{1}{8}$
27. $\frac{7}{9}-\frac{1}{8}$
28. $\frac{4}{5}-\frac{1}{3}$
29. $\frac{4}{8}+\frac{1}{4}$
30. $\frac{4}{5} of\ 90$
31. $\frac{5}{6}-\frac{1}{9}$
32. $\frac{5}{11}+\frac{3}{10}$
33. $\frac{1}{2}+\frac{1}{5}$
34. $\frac{4}{7}\div\frac{9}{10}$

35. $\frac{3}{11} \div 7$

36. $\frac{8}{9} of\ 63$

37. $\frac{4}{7} + \frac{1}{3}$

38. $\frac{5}{11} + \frac{2}{5}$

39. $\frac{1}{3} + \frac{1}{4}$

40. $\frac{2}{9} \times \frac{1}{2}$

41. $\frac{3}{4} of\ 52$

42. $\frac{9}{13} - \frac{3}{8}$

43. $\frac{2}{7} + \frac{1}{3}$

44. $\frac{6}{11} - \frac{2}{13}$

45. $\frac{3}{7} of\ 42$

46. $\frac{3}{8} + \frac{1}{7}$

47. $10 - \frac{4}{7}$

48. $15 \times \frac{5}{6}$

49. $8 \times \frac{7}{2}$

50. $\frac{4}{17} \div 3$

51. $\frac{9}{10} of\ 30$

52. $\frac{8}{9} \times \frac{9}{10}$

53. $\frac{4}{7} + \frac{1}{2}$

54. $\frac{8}{11} \div \frac{2}{7}$

55. $\frac{5}{6} of\ 72$

56. $\frac{5}{6} + \frac{5}{6}$

57. $\frac{1}{4} \times \frac{1}{5}$

58. $\frac{9}{10} \div \frac{3}{4}$

59. $\frac{2}{7} - \frac{1}{10}$

60. $\frac{2}{11} of\ 55$

Answers

1. $\frac{57}{40} \to 1\frac{17}{40}$
2. $\frac{154}{117} \to 1\frac{37}{117}$
3. $\frac{82}{360} \to \frac{41}{180}$
4. $\frac{7}{2} \to 3\frac{1}{2}$
5. $\frac{18}{15} \to \frac{6}{5} \to 1\frac{1}{5}$
6. $\frac{57}{108} \to \frac{19}{36}$
7. $\frac{14}{80} \to \frac{7}{40}$
8. $\frac{22}{21} \to 1\frac{1}{21}$
9. 22
10. $\frac{87}{10} \to 8\frac{7}{10}$
11. $\frac{26}{7} \to 3\frac{5}{7}$
12. $\frac{27}{40}$
13. $\frac{53}{247}$
14. $\frac{242}{360} \to \frac{121}{180}$
15. $\frac{1}{14}$
16. $\frac{6}{45} \to \frac{2}{15}$
17. $\frac{3}{180} \to \frac{1}{60}$
18. $\frac{34}{80} \to \frac{17}{40}$
19. $\frac{8}{21}$
20. 16
21. $\frac{13}{3} \to 4\frac{1}{3}$
22. 6
23. $\frac{10}{7} \to 1\frac{3}{7}$
24. $\frac{12}{36} \to \frac{1}{3}$
25. $\frac{5}{6}$
26. $\frac{65}{72}$
27. $\frac{47}{72}$
28. $\frac{7}{15}$
29. $\frac{24}{32} \to \frac{3}{4}$
30. 72
31. $\frac{39}{54} \to \frac{13}{18}$
32. $\frac{83}{110}$
33. $\frac{7}{10}$
34. $\frac{40}{63}$
35. $\frac{3}{77}$
36. 56
37. $\frac{19}{21}$
38. $\frac{47}{55}$

39. $\frac{7}{12}$

40. $\frac{2}{18} \rightarrow \frac{1}{9}$

41. 39

42. $\frac{33}{104}$

43. $\frac{13}{21}$

44. $\frac{56}{143}$

45. 18

46. $\frac{29}{56}$

47. $\frac{66}{7} \rightarrow 9\frac{3}{7}$

48. $\frac{75}{6} \rightarrow \frac{25}{2} \rightarrow 12\frac{1}{2}$

49. $\frac{56}{2} \rightarrow 28$

50. $\frac{4}{51}$

51. 27

52. $\frac{72}{90} \rightarrow \frac{4}{5}$

53. $\frac{15}{14} \rightarrow 1\frac{1}{14}$

54. $\frac{56}{22} \rightarrow \frac{28}{11} \rightarrow 2\frac{6}{11}$

55. 60

56. $\frac{10}{6} \rightarrow \frac{5}{3} \rightarrow 1\frac{2}{3}$

57. $\frac{1}{20}$

58. $\frac{36}{30} \rightarrow \frac{6}{5} \rightarrow 1\frac{1}{5}$

59. $\frac{13}{70}$

60. 10

The End.

Printed in Great Britain
by Amazon